Macmillan/McGraw-Hill Science

AUTHORS

Mary Atwater
The University of Georgia
Prentice Baptiste
University of Houston
Lucy Daniel
Rutherford County Schools
Jay Hackett
University of Northern Colorado
Richard Moyer
University of Michigan, Dearborn
Carol Takemoto
Los Angeles Unified School District
Nancy Wilson
Sacramento Unified School District

What do the dog, car, and driver have in common? They are all made of matter!

Macmillan/McGraw-Hill School Publishing Company
New York Chicago Columbus

CONSULTANTS

Assessment:

Janice M. Camplin
Curriculum Coordinator, Elementary Science
Mentor, Western New York
Lake Shore Central Schools
Angola, NY

Mary Hamm
Associate Professor
Department of Elementary Education
San Francisco State University
San Francisco, CA

Cognitive Development:

Dr. Elisabeth Charron
Assistant Professor of Science Education
Montana State University
Bozeman, MT

Sue Teele
Director of Education Extension
University of California, Riverside
Riverside, CA

Cooperative Learning:

Harold Pratt
Executive Director of Curriculum
Jefferson County Public Schools
Golden, CO

Earth Science:

Thomas A. Davies
Research Scientist
The University of Texas
Austin, TX

David G. Futch
Associate Professor of Biology
San Diego State University
San Diego, CA

Dr. Shadia Rifai Habbal
Harvard-Smithsonian Center for Astrophysics
Cambridge, MA

Tom Murphree, Ph.D.
Global Systems Studies
Monterey, CA

Suzanne O'Connell
Assistant Professor
Wesleyan University
Middletown, CT

Environmental Education:

Cheryl Charles, Ph.D.
Executive Director
Project Wild
Boulder, CO

Gifted:

Sandra N. Kaplan
Associate Director, National/State Leadership
Training Institute on the Gifted/Talented
Ventura County Superintendent of Schools Office
Northridge, CA

Global Education:

M. Eugene Gilliom
Professor of Social Studies and Global Education
The Ohio State University
Columbus, OH

Merry M. Merryfield
Assistant Professor of Social Studies and Global Education
The Ohio State University
Columbus, OH

Intermediate Specialist

Sharon L. Strating
Missouri State Teacher of the Year
Northwest Missouri State University
Marysville, MO

Life Science:

Carl D. Barrentine
Associate Professor of Biology
California State University
Bakersfield, CA

V.L. Holland
Professor and Chair, Biological Sciences Department
California Polytechnic State University
San Luis Obispo, CA

Donald C. Lisowy
Education Specialist
New York, NY

Dan B. Walker
Associate Dean for Science Education and Professor of Biology
San Jose State University
San Jose, CA

Literature:

Dr. Donna E. Norton
Texas A&M University
College Station, TX

Tina Thoburn, Ed.D.
President
Thoburn Educational Enterprises, Inc.
Ligonier, PA

Macmillan/McGraw-Hill School Division
10 Union Square East
New York, New York 10003

Printed in the United States of America

ISBN 0-02-274266-2 / 4

3 4 5 6 7 8 9 VHJ 99 98 97 96 95 94 93

Monument Valley, Utah

Mathematics:

Martin L. Johnson
Professor, Mathematics Education
University of Maryland at College Park
College Park, MD

Physical Science:

Max Diem, Ph.D.
Professor of Chemistry
City University of New York, Hunter College
New York, NY

Gretchen M. Gillis
Geologist
Maxus Exploration Company
Dallas, TX

Wendell H. Potter
Associate Professor of Physics
Department of Physics
University of California, Davis
Davis, CA

Claudia K. Viehland
Educational Consultant, Chemist
Sigma Chemical Company
St. Louis, MO

Reading:

Jean Wallace Gillet
Reading Teacher
Charlottesville Public Schools
Charlottesville, VA

Charles Temple, Ph. D.
Associate Professor of Education
Hobart and William Smith Colleges
Geneva, NY

Safety:

Janice Sutkus
Program Manager: Education
National Safety Council
Chicago, IL

Science Technology and Society (STS):

William C. Kyle, Jr.
Director, School Mathematics and Science Center
Purdue University
West Lafayette, IN

Social Studies:

Mary A. McFarland
Instructional Coordinator of
Social Studies, K-12, and
Director of Staff Development
Parkway School District
St. Louis, MO

Students Acquiring English:

Mrs. Bronwyn G. Frederick, M.A.
Bilingual Teacher
Pomona Unified School District
Pomona, CA

Misconceptions:

Dr. Charles W. Anderson
Michigan State University
East Lansing, MI

Dr. Edward L. Smith
Michigan State University
East Lansing, MI

Multicultural:

Bernard L. Charles
Senior Vice President
Quality Education for Minorities Network
Washington, DC

Cheryl Willis Hudson
Graphic Designer and Publishing Consultant
Part Owner and Publisher, Just Us Books, Inc.
Orange, NJ

Paul B. Janeczko
Poet
Hebron, MA

James R. Murphy
Math Teacher
La Guardia High School
New York, NY

Ramon L. Santiago
Professor of Education and Director of ESL
Lehman College, City University of New York
Bronx, NY

Clifford E. Trafzer
Professor and Chair, Ethnic Studies
University of California, Riverside
Riverside, CA

STUDENT ACTIVITY TESTERS

Jennifer Kildow
Brooke Straub
Cassie Zistl
Betsy McKeown
Seth McLaughlin
Max Berry
Wayne Henderson

FIELD TEST TEACHERS

Sharon Ervin
San Pablo Elementary School
Jacksonville, FL

Michelle Gallaway
Indianapolis Public School #44
Indianapolis, IN

Kathryn Gallman
#7 School
Rochester, NY

Karla McBride
#44 School
Rochester, NY

Diane Pease
Leopold Elementary
Madison, WI

Kathy Perez
Martin Luther King Elementary
Jacksonville, FL

Ralph Stamler
Thoreau School
Madison, WI

Joanne Stern
Hilltop Elementary School
Glen Burnie, MD

Janet Young
Indianapolis Public School #90
Indianapolis, IN

CONTRIBUTING WRITER

Linda Barr

Properties of Matter

Lessons **Themes**

Activities!

EXPLORE

TRY THIS

Features

Literature Links

Social Studies Link

Art Links

Health Links

Focus on Environment

Focus on Technology

Departments

Properties of Matter

People are constantly changing their ideas as they learn new things. For example, philosophers used to think everything in the universe was made of four elements—air, fire, soil, and water. Now we know everything is actually made of 92 naturally occurring elements. All of these elements are different forms of matter. Did you know that?

Minds On! To start your exploration of matter, see if you can answer these questions in your ***Activity Log*** on page 1.

- What is between you and the ceiling?
- How many different materials is your desk made of?
- What is the smallest thing in your desk?
- Is anything in your desk moving? What?
- Can you make something in your desk move faster? If so, how? ●

People are still learning about matter. Because of what we've learned, we have been able to send shuttles into outer space and cure deadly diseases. We have also changed the kinds of clothes we wear and the kinds of food we eat.

In this unit, you'll do activities that will help you discover and explore matter around you. You'll be able to understand the changes you see every day, from ice cubes melting in your glass to a balloon rising in the air.

In everyday life, we learn to expect certain things. What do you expect when you pour chocolate syrup into milk? When you fall on the sidewalk, what do you expect to happen when your knee hits the hard cement? Ouch!

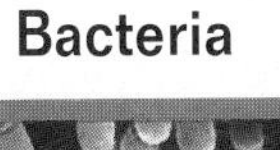

Bacteria

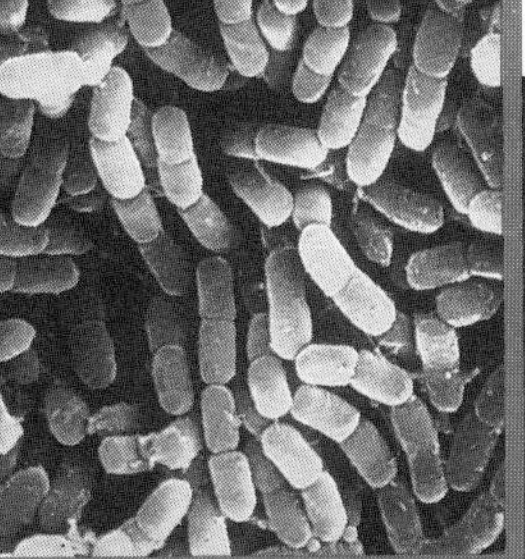

Hot air balloon

Chocolate syrup and milk

Space shuttle blasting off

People doing research often expect certain things to happen, too. But experiments don't always turn out as expected. About 50 years ago, a young scientist named Roy Plunkett was experimenting with different kinds of matter. He expected to invent a new kind of gas to be used to cool things. But he didn't get the results he expected.

Rather than treat the experiment as a failure, Roy Plunkett carefully examined what did happen. As a result, he made a discovery that's now used all over the world—in automobile motors, in artificial heart valves, in desert tents, and probably in your kitchen! At the end of this unit, you'll find out what he discovered. In the meantime, as you learn how matter acts and interacts, you'll make some discoveries that might surprise you, too.

Roy Plunkett

Made With Peanuts!

CAREERS

A scientist named George Washington Carver used his knowledge of matter in experiments with peanuts. His understanding of how matter interacts with other matter helped him invent many uses for the peanut. He developed more than 300 products, including shampoo, paper, cheese, and ice cream—all made from the peanut!

Johnnie H. Watts was inspired by the work of George Washington Carver. Dr. Watts received an advanced degree from the University of Chicago, where she studied many areas of science and math. She is using her knowledge of the interactions of matter to continue Mr. Carver's study of food. As a food chemist, Dr. Watts is learning more about the ways food keeps us healthy. In particular, Dr. Watts has studied a nutrition problem in young adults.

George Washington Carver with students in his laboratory

Science in Literature Literature Link

Books are not only made of matter, but also tell you about matter and make you think about matter in new ways.

Sugaring Time, by Kathryn Lasky. New York: Macmillan Publishing Company, 1983.

The time of year when winter gives way to spring is "sugaring time" in central Vermont. It is during this season that the sap begins to flow in the maple trees and the Lacey family makes maple syrup.

The Laceys use their knowledge of maple trees to place the taps and take the correct amount of sap without damaging the trees. They then use their understanding of the properties of the sap to change it into delicious syrup.

***Building a House*, by Ken Robbins. New York: Four Winds Press, 1984.**

Have you ever wondered how your house or apartment was built? Building a house takes many people, lots of time, and plenty of matter, such as concrete, wood, glass, electrical wires, paint, and much more. This book follows architects as they make drawings and models of a family's dream house. It shows masons laying a foundation of concrete and carpenters building the frame and roof with wood and nails. Then electricians and plumbers put lights and running water into the house. Step by step, *Building a House* explains the mysteries of turning an architect's ideas into a family's home.

Other Good Books To Read

***Fun With Physics*, by Susan McGrath. Washington, D.C.: The National Geographic Society, 1986.**

Susan McGrath uses examples from our everyday world to explain physics and make it fun.

***Ice Cream*, by William Jaspersohn. New York: Macmillan, 1988.**

This book shows how much matter goes into making ice cream, and how it ends up in the store ready for you to buy.

***The Mud Pony*, by Caron Lee Cohen. New York: Scholastic Inc., 1989.**

The Mud Pony is the story of a Native-American boy and his adventures with the pony he makes out of mud.

Identifying Matter

Can you think of at least one way a car and a dog are different? Of course you can. By thinking of the properties of each, you can probably come up with many differences. But can you think of a way in which they are the same?

It's easy to see ways in which the cars pictured here are the same. The dogs pictured here look the same, too. But in what way are cars, dogs, and people the same? One answer is they're all made of matter, just like you. In this lesson you'll learn more about matter, and about the ways we identify different kinds of matter. To practice identifying matter, do the Try This Activity on page 13.

Everything in the world is made of matter. Matter includes solid things like rocks, liquids like milk, and gases like the air you're breathing. It is easy to identify some kinds of matter by a quick description.

TRY THIS

Activity!

What Am I?

What You Need

Activity Log **page 2**

Can you describe something so completely that another person can identify it? What information will help the other person visualize the object you are explaining? Think of three objects in your bedroom. On a piece of paper, describe each object without naming it. Use four features to describe each object. Exchange papers with a classmate. In your ***Activity Log,*** identify and draw the object your classmate described.

But sometimes a more definite identification is needed. If you describe a person as being a female, of medium height, with brown curly hair and brown eyes, that description will fit many women. What about identical twins? Everything about them seems to be the same. But there are characteristics that are different for everyone, even twins. These characteristics can provide a definite identification for a specific person. Can you think of one of these characteristics?

EXPLORE

Activity!

Who Is It?

One way to identify a specific person is by a characteristic everyone has—fingerprints. Everyone has fingerprints that are different from everyone else's. Though fingerprints are different, the features that make up fingerprints can be put into three big groups.

In this activity you are going to observe and classify your own fingerprints. You will then use the classification of fingerprints to solve a mystery!

What You Need

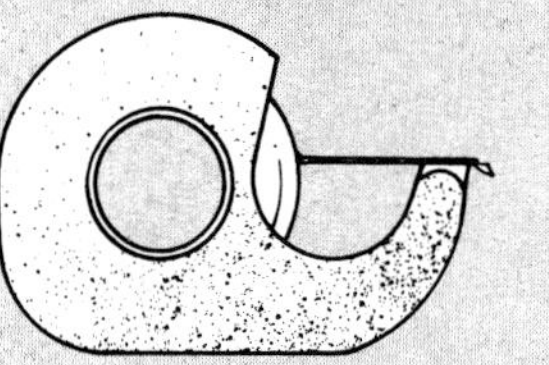
transparent tape

scrap paper

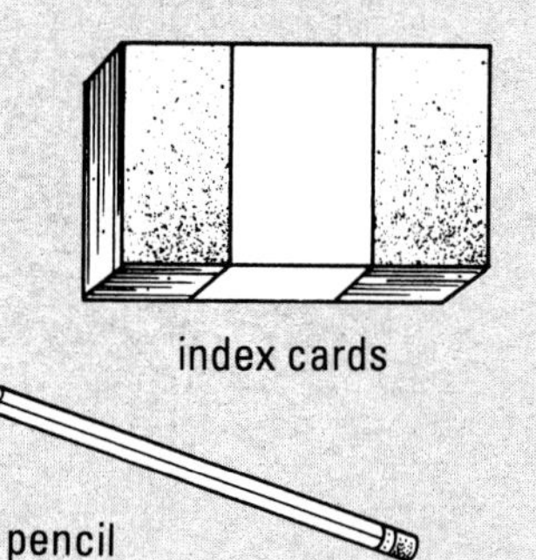
index cards

pencil

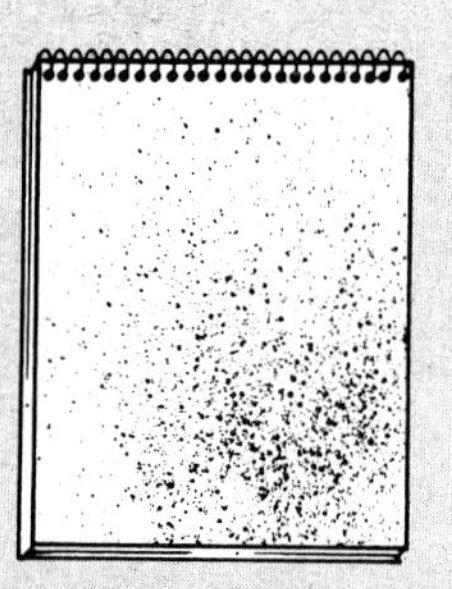
Activity Log page 3-4

What To Do

1 Rub the pencil on a scrap of paper to make a dark black smudge. Rub your right thumb in the pencil smudge. Your thumb should be darkened by the pencil lead.

2 Have your partner put a piece of tape around your thumb, taking it off right away, and sticking it on one of the 3 x 5 in. cards. What do you see? Write your name on this card and keep it.

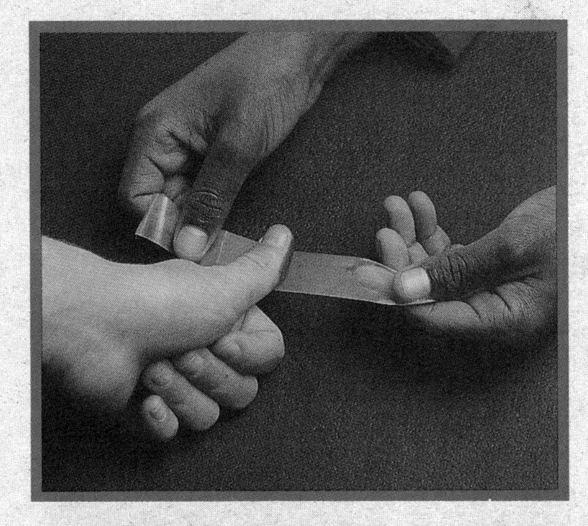

3 Each person in your group should make one more copy of his or her thumbprint on a separate 3 x 5 in. card. Write your name on the back of the card and give it to your teacher.

4 Carefully look at the prints in your group. How many ways can you classify their properties?

5 Predict which property is the most common among the prints in your group.

What Happened?

1. What type of print do you have? Check to see if everyone in your group agrees with how to classify prints.
2. Which type is most common in your group? In your whole class?
3. How did your results compare with your prediction?

What Now?

1. How could you use fingerprints to identify someone?
2. Examine the print on the card your teacher gave you and identify the person in your group it belongs to. Turn the card over to see if you are correct.

Using Properties

In the Explore Activity, you discovered that fingerprints are different, though they have some similar characteristics. You might know more than one female of medium height, with brown curly hair and brown eyes. However, as with fingerprints, the unique way these and other characteristics come together makes each person different from everyone else.

A **property** (prop′ ər tē) is a characteristic of an object. It is often the way something looks, smells, tastes, or feels. In the Explore Activity on pages 14–15, you discovered the way properties of fingerprints can be used to identify people. Think back to the Try This Activity on page 13. What properties did you use to describe the objects in your bedroom? Were there other properties that would have helped to identify each object?

Properties can be used to classify objects. They can also be used to identify objects. For example, you can use your knowledge of properties to identify what is on the pizza shown here.

n addition to fingerprints, the pattern of the blood vessels located at the back of your eyes can be used to identify you. Each person has a unique pattern of blood vessels.

The special machine pictured above uses light to code the pattern of the blood vessels into eyeprints. The code provides a very accurate method of identification. If eyeprints for both eyes are used, a mistake in identification is made only about one time in a trillion!

Your voice is another unique property that can be used to identify you. Your voiceprint remains the same even if you try to change your voice or if you have a cold. To make a voiceprint, you say a password into a special recorder. A computer records characteristics of your voice in code on a plastic card about the size of a credit card.

Minds On! In your ***Activity Log*** on page 5, write or draw a plan for a way that fingerprints, eyeprints, or voiceprints might be used. For example, do you know anyone who has ever been locked out of his or her car or home? How would a special method of identifying that person have helped?

An example of a printout of a human voiceprint

We have been thinking about and using some properties of matter, but what is matter? **Matter** is anything that takes up space and has mass. Matter also has properties by which it can be described.

Every day we use properties to identify matter. For example, have you ever asked for a new notebook for school? You probably described it by some of its properties—three rings, 8.5 inches by 11 inches, red or blue. These properties relate to size and color. When you tried to put the notebook into your desk, you might have noticed another property of matter.

This child's notebook doesn't fit into her desk. She has learned that two objects cannot occupy the same space at the same time.

Minds On! Put your right index finger on your desk. Without lifting your finger, put your left index finger in exactly the same space. Can you do it? Now try to put two books in the same place. What do these two quick activities tell you? You've just proved that each finger and each book is made of matter and takes up space.●

Understanding the properties of matter is necessary in order to choose the best materials for a specific project. For example, if you wanted a kite to last forever, would you make it out of steel? Do the Try This Activity on this page to choose the best material for a project.

The amount of matter in an object is the object's **mass.** A house has much more mass than a book.

What has more mass than your pencil? What has less mass than your pencil? Does your pencil gain or lose mass as you sharpen it? Why?

A brick has more mass than a basketball.

TRY THIS Activity!

What Good Are Properties?

What You Need

2 plastic drinking straws, 3 pieces of paper, piece of cardboard, tape, metric ruler, various small objects to use as weights, 2 textbooks, *Activity Log* page 6

Place the two books 15 cm apart on your desk. Use the materials, one at a time, to make a bridge across the space between the books. Predict which bridge will be the strongest. Was your prediction correct? How does this relate to the properties of matter? Experiment with different shapes of bridges made of paper. In your ***Activity Log,*** draw a picture of the paper bridge design that was the strongest.

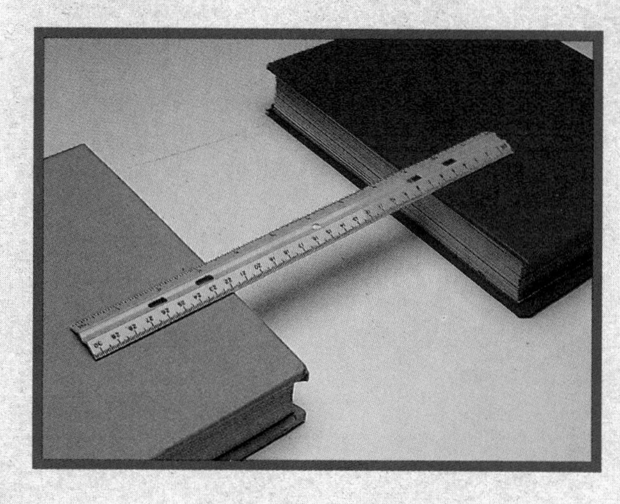

Plastics, the Second Time Around

Focus on Environment

If you throw away objects made from plastic, where do they go? Some communities take trash to landfills. But we have too much trash and too little room to continue to create landfills. Some communities have recycling programs. Objects made of plastic, along with paper, glass, and metals, are collected and processed to be reused. Plastic detergent jugs, soft drink bottles, and milk containers can be shredded into flakes, which are then cleaned and made into new products.

Your comb might have been an empty detergent jug.

This "wooden" bench may have been hundreds of used plastic milk containers.

This carpeting may have been more than one thousand old soft drink bottles.

Does your community recycle? If not, you may wish to write to your mayor, or another local government official, to find out how such a program could be started. You can make a valuable contribution to saving our environment.

Using Properties To Choose Materials

Music/Art Link

Artists must understand the properties of their materials to choose the right ones for each project.

Create a poster using various art materials. As you make the poster, in your ***Activity Log*** on page 7 list two properties of the materials you use. Give an example of a project in which each material would be good to use, and one project in which the material would not be appropriate.

Sum It Up

You and everything you touch is matter, which can be identified by properties.

Minds On! Review the definition of matter. Make a list of five things that are matter. Now make a list of five things that are not matter. Compare your list with a classmate's list. ●

Your desk is hard, probably light brown or gray, and probably shaped like a rectangle. Your own properties are much more complex than those of the desk. Your fingerprints, eyeprints, and even your voiceprint identify you as a unique bit of matter in this amazing universe.

Critical Thinking

1. Why can't we use a football to play basketball? What properties of a football would ruin a basketball game?

2. What properties of a beach ball make it fun in the water, but hard to use for bowling?

3. How does the mass of a ball affect how far you can throw it?

Theme T SCALE and STRUCTURE

You have enough players for two baseball teams, a bat, a ball, and an unmarked field. How can you measure the field so the bases are all the same distance apart? Do you need to have the correct measurements in order to play?

AMOUNTS COUNT

In the last lesson, you learned that everything in the universe is made of matter, including cars, dogs, and you. You also learned about some properties of matter—size, shape, color, mass. Matter has another property—it can be measured. People have not always had rulers, measuring tapes, or scales, but they have always had the need to measure. One way people have measured without the use of a measuring tool was by using body parts.

Music/Art Link

Discover Proportions Da Vinci Used

Trace your hand and forearm on a piece of craft paper. How many finger widths does it take to cover your palm? How many palms is the distance from your elbow to your fingertip? Record your answers on page 8 of your *Activity Log.*

Five hundred years ago, the Italian artist Leonardo da Vinci (lē′ ə när′ dō də vin′ chē) studied the relationships of body parts. He found that a person's palm is equal to four widths of that person's finger. Six palms equals one cubit. Understanding the relationships of body parts and using nonstandard ways to measure helped Da Vinci draw bodies with correct proportions. Have you ever used a nonstandard method of measuring?

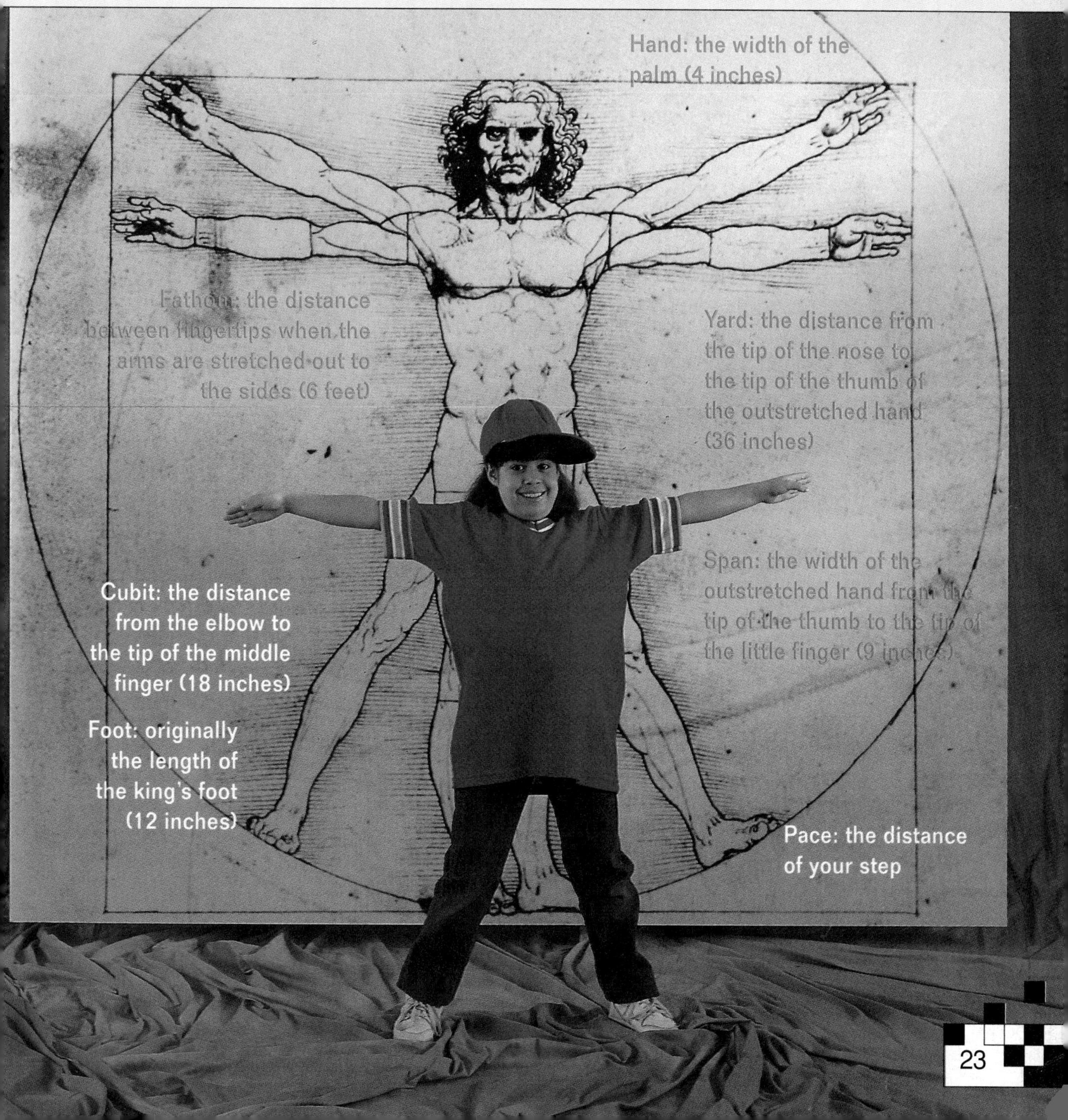

Hand: the width of the palm (4 inches)

Fathom: the distance between fingertips when the arms are stretched out to the sides (6 feet)

Yard: the distance from the tip of the nose to the tip of the thumb of the outstretched hand (36 inches)

Span: the width of the outstretched hand from the tip of the thumb to the tip of the little finger (9 inches)

Cubit: the distance from the elbow to the tip of the middle finger (18 inches)

Foot: originally the length of the king's foot (12 inches)

Pace: the distance of your step

EXPLORE

Activity!

How Can Matter Be Measured?

In this activity you will be using both nonstandard and standard units to measure the same objects. How do you think the measurements you and your classmates make will compare?

What You Need

Activity Log page 9-10

meter tape

What To Do

1 Predict the length of your desk in spans. Now measure the length of your desk in spans. Write the number in your *Activity Log.*

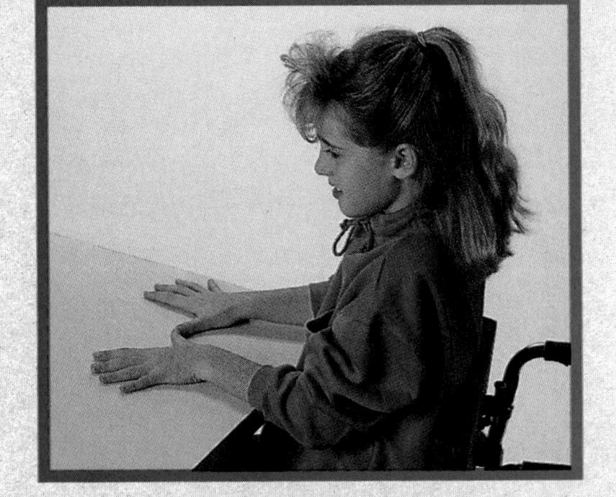

2 Predict the length of your desk in centimeters. Measure and record the length in your *Activity Log.*

3 Predict the length of your classroom in paces. Measure the length of the room in your paces and record the measurement in your *Activity Log.*

4 Predict the length of your classroom in meters. Measure and record the length in your ***Activity Log.***

5 Measure the length of your desk in millimeters. How does this measurement compare with the measurement of the length of your desk in centimeters? Record your findings in your ***Activity Log.***

6 Measure the length of your classroom in centimeters. Compare this figure with the measurement in meters. Record your findings in your ***Activity Log.***

What Happened?

1. Compare your measurements in spans and paces with those of your classmates.
2. How did your measurements in centimeters and meters compare with those of your classmates?
3. What did you notice about the measurement of your desk in millimeters, centimeters, and meters?

What Now?

1. If giant submarine sandwiches cost $2.00 per span, would you rather have the person with the shortest or longest span measure your portion? Why?
2. Whose span do you think the sandwich shop owner would rather use? Why?
3. Most of the time, we use standard units of measurement rather than parts of the body to measure objects. Why?
4. When are nonstandard units of measurement useful?

Measuring Matter

As you can see, even buying submarine sandwiches would be difficult without a standard unit of measurement. In fact, many areas of our lives would change if we couldn't agree on how to measure things. We would live in a world of mismatched parts.

To avoid this problem, almost all countries use the metric system as a standard way to measure. This system is based on the number 10. That means we multiply one unit by 10 (or 100 or 1,000) to get a larger unit. We divide a unit by 10 (or 100 or 1,000) to get a smaller unit. You can use the metric system to measure the physical properties of matter.

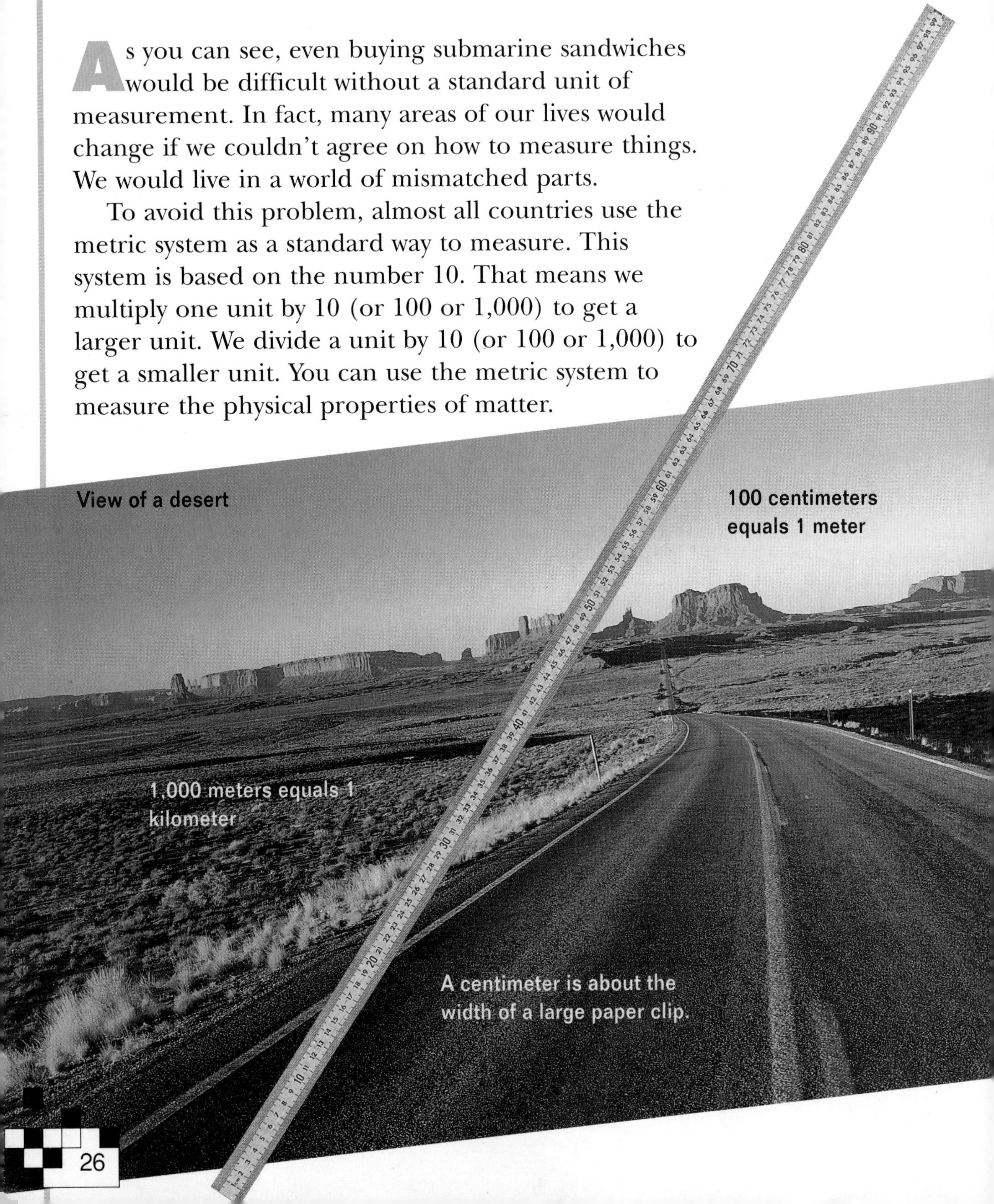

View of a desert

100 centimeters equals 1 meter

1,000 meters equals 1 kilometer

A centimeter is about the width of a large paper clip.

How Much Farther?

Can you imagine how difficult it would be to measure the distance between two cities using a ruler? We're lucky we don't have to. Map-makers help us measure distances by producing road maps like the one below. Find Northfield in square A1, Centerville in B2, and Clark in C3. The map-maker has put a scale on the map to show us that 1.0 centimeters on the map is equal to 1 kilometer on the highway. Rather than using a ruler on the highway, use it on the map to find the distance in kilometers between Northfield and Centerville. Next, find the distance between Northfield and Clark. Can you think of another way to measure these distances? If you had to use a different measuring tool, what would you use? If you didn't have any measuring tools, what could you use? Use page 11 of your ***Activity Log*** to record your findings.

Volume

Volume (vol′ ūm) is the amount of space that something occupies. All matter has volume because all matter takes up space. Volume is also a property that can be measured. The volume of a rectangular object is calculated by multiplying the length times the width times the height.

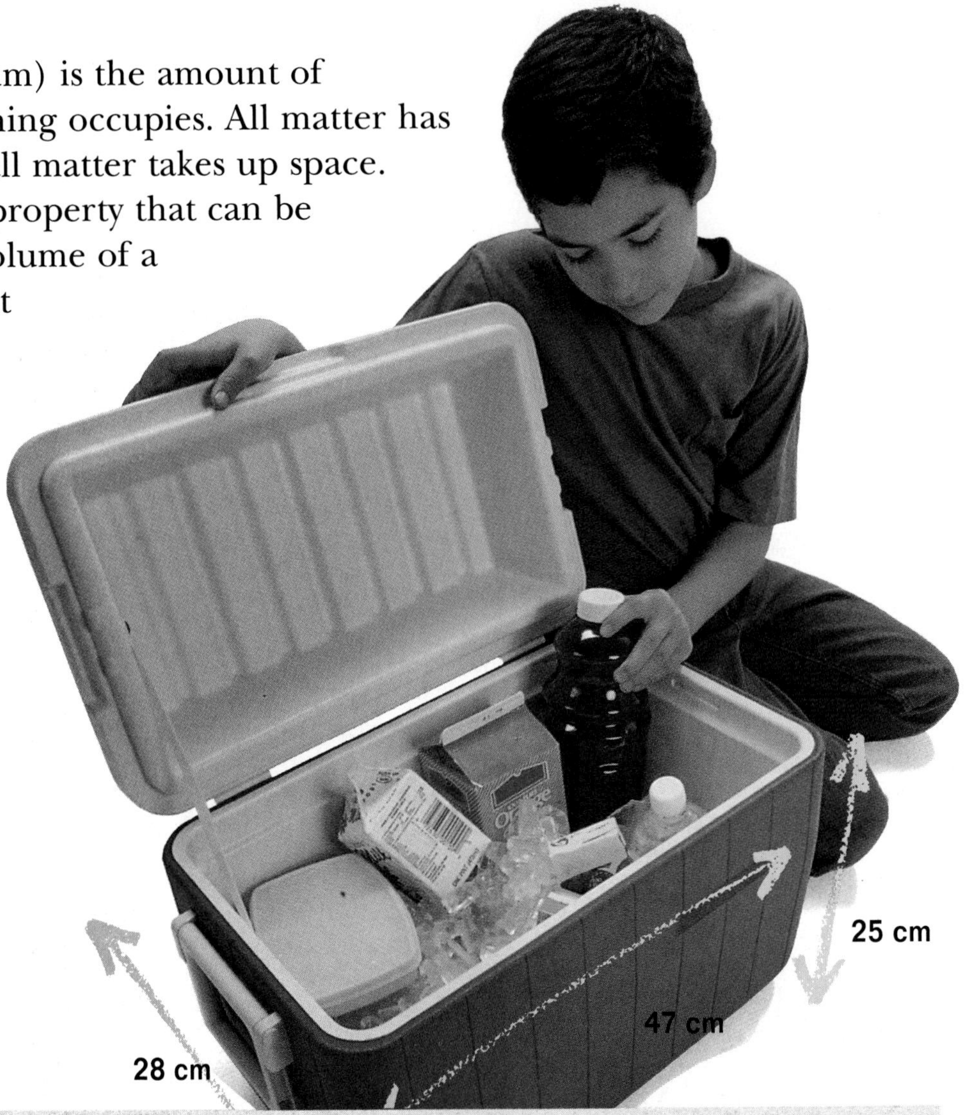

Finding room for a two-liter jar can be a problem. A juice box easily fits in a cooler even when it looks full. The two-liter jar has more height, width, and depth. In other words, it has much more volume.

25 cm

47 cm

28 cm

TRY THIS Activity!

What's the Volume?

What You Need

meter tape, *Activity Log* page 12

First, using a calculator, find the volume of the cooler pictured on this page. Next, find the volumes of some rectangular objects in your classroom. You could measure a book, a box, an aquarium, or an animal cage. What unit of measurement would you use for your calculations? What unit would you use for your measurements? What unit would be better if you wanted to measure the room?

An eraser might be 7 centimeters long, 3 centimeters wide, and 1 centimeter high. Multiply these numbers to get the volume of the eraser: 7 centimeters × 3 centimeters × 1 centimeter = 21 cubic centimeters.

TRY THIS

Activity!

Change the Shape, Change the Volume?

What You Need

clay, metric ruler, 50-mL graduated cylinder, water, *Activity Log* page 13

Does the volume of an object change when you change its shape? Try to shape a piece of clay into a cube that is exactly 1 cm on each side. What is the volume of the clay?

Once you are sure that your cube measures 1 cm on each side, fill a 50-mL graduated cylinder to the 25-mL mark with water. Drop your cube of clay into the 25 mL of water. Read the new water level on the graduated cylinder.

Record the change in the water level in your ***Activity Log.*** How did this change in water level compare with the measurements of your cube?

Now, change your cube of clay into a new shape. Drop this new shape into 25 mL of water in the graduated cylinder. Record this change in the water level in your ***Activity Log.***

How did the water level change for the new shape? What can you conclude about how a change of shape affects volume?

Mass

How do a golf ball and a table tennis ball compare? A golf ball and a table tennis ball are nearly the same size. However, a golf ball has more matter than a table tennis ball. The amount of matter in an object is the mass of the object. Mass is measured in grams or kilograms.

Mass not only differs from volume, but it also differs from weight. Mass and weight are often confused.

Weight is a measure of the pull of gravity on an object. Look at the illustration of the astronaut on Earth, on the moon, and on Jupiter. You can see that the mass of the astronaut does not change from planet to planet. However, because of the differences in the pull of gravity among the planets, the astronaut's weight does change.

Do the Try This Activity on page 31 to practice finding the mass and weight of an object. The metric unit for weight is the **newton,** which you will use in the activity.

180 kg

4,800 N

Literature Link

Sugaring Time

As you read *Sugaring Time,* think about how many properties of the sap are measured in order to know when to stop the process at the correct time to get great syrup. Were you surprised at how many gallons of sap it takes to make one gallon of syrup?

Jupiter

TRY THIS

Activity!

Mass vs. Weight

What You Need

double-pan balance, spring scale, standard masses, cassette tape, *Activity Log* page 14

How can both the mass and the weight of an object be measured? Place a cassette tape on one pan of a double-pan balance. Add standard masses to the other pan until it balances. The sum of the standard masses equals the mass of the cassette tape. To weigh the cassette, hook it to a spring scale. Read the pointer on the scale to find the weight in newtons. Record both the mass and the weight of the cassette in your ***Activity Log.*** If you were to do this activity on the moon, which would change—mass or weight? Explain your answer.

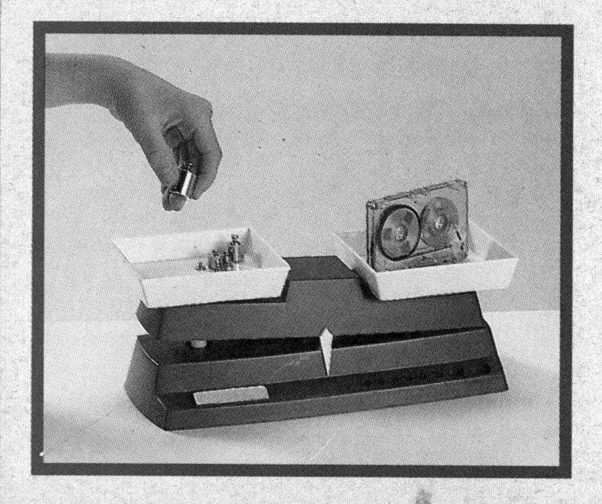

Density

Imagine you ordered regular golf balls and plastic golf balls from a sporting goods company. If both kinds of balls came in the mail in the same-size box, how could you tell them apart? The size of the box wouldn't help because both boxes have the same volume. However, the masses of the balls in the boxes would be different. The box with the regular golf balls would have the larger mass. Since regular golf balls have more mass than plastic golf balls even though both kinds of balls are the same size, we say the density of the regular golf balls is larger.

Objects that sink in water, like regular golf balls, have a density that's greater than the density of water.

Suppose you dropped plastic golf balls and regular golf balls into water. Which ones do you think would sink? Often the density of objects is compared by using water.

Objects that float in water, like plastic golf balls, have a density that's less than the density of water.

TRY THIS **Activity!**

Will It Sink or Will It Float?

What You Need

1 cup filled with unpopped popcorn, 1 cup filled with popped popcorn, large clear plastic cup, pan balance, water, *Activity Log* page 15

How do the densities of popped and unpopped popcorn compare? Put the cup of unpopped popcorn on one pan of the pan balance and the cup of popped popcorn on the other pan. Record which cup has the greater mass in your ***Activity Log.*** Now, fill the large clear plastic cup with water. Pour some of the popped and some of the unpopped corn into the water. Which one is more dense? Did the mass change from unpopped to popped? Did the volume change? If the volume of an object stays the same but the mass increases, what happens to the density? Why? Record your observations in your ***Activity Log.***

Measurements Count

Knowing the relationships between units of standard or nonstandard measure helps you solve some everyday problems more quickly. For example, Da Vinci studied the relationships of the measurements of body parts, and that information is still being used today. Clothing makers use relationships of body parts to make clothing in the correct proportions. Furniture makers can make comfortable chairs using similar information.

Minds On! Look back at the picture Leonardo da Vinci drew of the human body on page 23. Da Vinci observed that a person's fathom is equal to another of his or her measurements. Can you guess what measurement it is by looking at the picture? ●

Building a House Literature Link

Read *Building a House* by Ken Robbins to see many examples of people depending on exact measurements to complete their jobs. For instance, in order for the person to put the windows in the new home, the carpenter must cut the openings the correct size.

Does a tree house have to be measured perfectly for it to serve its purpose? It probably does not matter if it is a little wider in some places than in others.

Sum It Up

It's important to understand not only the properties of matter, but also ways to measure those properties. Sometimes, using whatever you have handy, such as a stick or your arms, works just fine. At other times, a more specific weight, volume, or length is necessary.

Think about how different our lives would be if we didn't have a standard way to measure. How would a fast-food restaurant order enough hamburger? How would the correct amount of medicine be put into capsules? How could we have cars or jet planes or space shuttles—or even birthday cakes?

Accurate measurements are much more important when building a house, so that there aren't any gaps in the walls or ceiling.

Critical Thinking

1. You and your mother are going shopping for a pair of shoes for your little sister. How might you determine the correct size of shoes to purchase?

2. A flower bed 560 centimeters long is also 5.6 meters long. A pencil 7 inches long is also about 0.58 feet long. How is using the metric system easier than using customary units such as inches and feet?

3. If you double the volume of a substance and its mass doubles as well, what happens to its density?

Atoms, the Building Blocks of Matter

Do bricks look much like bananas or bicycles? No way! What gives bricks, bananas, and bicycles such different properties?

Look at the photos on the next page. Can you tell what they show?

What is the wall made of? Bricks, of course! You can see them. Bricks are the building blocks of the wall. Now look closely at the banana. What do you think makes up a banana? Does the banana have building blocks, too? What about the bicycle? Can you think of anything bricks, bananas, and bicycles have in common? Think about color, shape, size, hardness, or roughness.

The properties of bricks, bananas, and bicycles come from different combinations of the tiny building blocks of matter called atoms. In this lesson you'll learn how bricks, bananas, bicycles, and you are made of many different kinds of atoms.

What is the smallest part of everything in these pictures, including the air?

Minds On! Has anyone ever cut an apple in half for you? Imagine that this person had a super-sharp knife and continued cutting the apple into smaller and smaller pieces. Is there an end to how many times the apple could be cut?

EXPLORE

Activity!

What Is Matter Made Of?

In this activity you'll look at some everyday objects and try to see what different types of matter they are made of.

What You Need

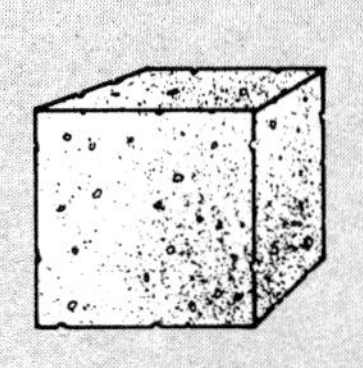

sugar cube

magazine photo

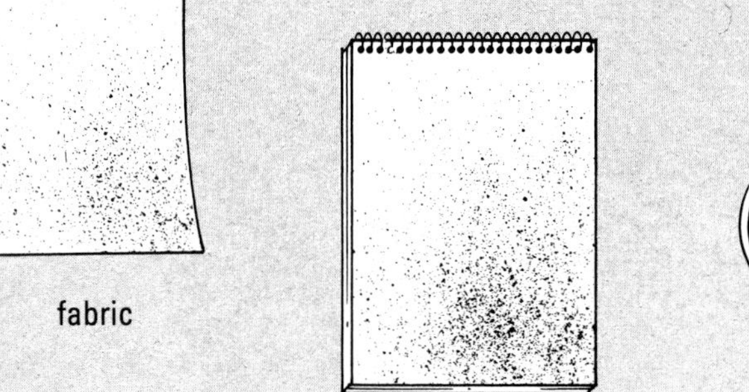

fabric

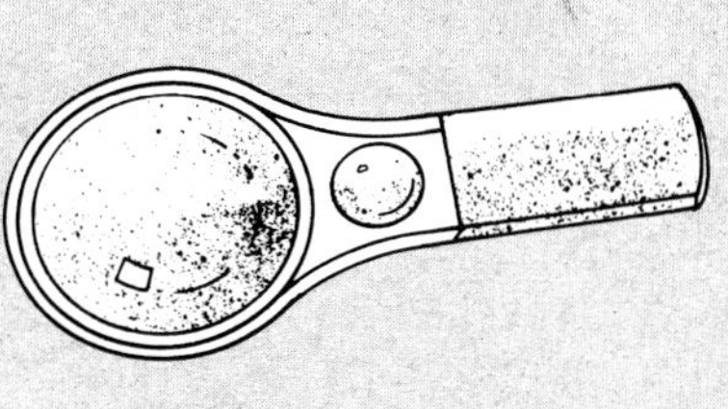

hand lens

***Activity Log* page 16-17**

What To Do

1 Look closely at a small portion of the picture. Draw or write a description of what you see in your ***Activity Log.*** Examine the same portion of the picture with the hand lens. Draw or write a description of what you see in your ***Activity Log.***

2 Look closely at the fabric and draw or write a description of what you see in your ***Activity Log.*** Examine the piece of fabric with the hand lens. Draw or write a description of what you see in your ***Activity Log.***

3 Look closely at the sugar cube. Draw or write a description of what you see in your ***Activity Log.*** Examine the sugar cube with the hand lens. Draw or write a description of what you see in your ***Activity Log.***

What Happened?

1. For each item, how is what you saw with the hand lens different from what you saw without it?
2. What smaller parts were each of the items you looked at made from?

What Now?

1. If you examined each of these items under a microscope, what would you expect to see?
2. If more powerful microscopes were available, what would you expect to see?

Atoms and Elements

As you saw in the Explore Activity, matter is made of smaller parts of the same matter. About 400 B.C., a Greek philosopher named Democritus (də mä′ kri tus′) developed the idea that matter could be divided into smaller and smaller pieces until it could no longer be divided. The word for the smallest piece of matter, *atom,* comes from the Greek word *atomos,* meaning "uncuttable." Democritus never saw an atom. He developed his ideas by thinking about something he could not see. Like philosophers, scientists often make discoveries by thinking about things they cannot see.

Atoms are the building blocks of matter. An **atom** (at′ əm) is the smallest piece of an element that still has the properties of that element. An **element** (el′ ə mənt) is made of only one kind of atom. If you could keep cutting an element in half again and again, you would finally get to the point where you couldn't cut it in half anymore and still have that element. At that point, you would have an atom.

A nickel is made of 25% of the element nickel and 75% of the element copper.

TRY THIS Activity!

Let's Get Physical

What You Need

miscellaneous objects, *Activity Log* page 18

Notice the physical properties of each item. Classify, or group, the items into three groups according to their properties. Which properties did you use? Did a classmate classify the items differently? Is one classification better than another? Explain your answer.

The most abundant elements on Earth are iron, oxygen, magnesium, silicon, nickel, and sulfur. You might be familiar with the names of some other elements, such as lead, aluminum, copper, tin, silver, and gold. These elements are metals.

Carbon combines with other elements to make substances as different as plastics and gasoline.

The element carbon is found in every living thing. When we burn wood or paper (which used to be living, of course), the flame gives off a black soot, which is carbon.

As Good As Gold

GLOBAL VIEW

It's not too late to join the gold rush! The search for gold has appealed to people of many backgrounds for over 100 years. In fact, half of the miners in California between 1848 and 1850 were Native Americans! Only a small amount of the existing gold was found during the gold rush. The rest, billions and billions of dollars' worth, is still in gold fields all over the world.

In Toronto, the windows of some buildings are coated with a thin film of material containing gold to keep heat outside in the summer and inside in the winter.

Many years ago, Carl Fabergé (fab uhr zhay') created a collection of golden eggs for the family of a Russian ruler.

A thin layer of gold on the visor of this helmet helps insulate the astronaut from the sun.

Shortly after the 1848 gold rush, explorers found gold in Australia.

Gold has always been valuable because it is so rare, so beautiful, and so useful.

Not all gold ends up in wedding bands and bracelets. Gold is used as money all over the world. All countries accept gold as payment from other countries. Scientists have found surprising uses for gold because of its physical properties. Gold does not rust or tarnish, and it allows electricity to flow easily. That's why communication cables on the floor of the ocean contain gold.

Gold also stretches. An ounce of gold can be stretched into an extremely thin wire 50 miles long. If you have a TV, VCR, computer, or calculator, it probably contains some thin gold wires.

Gold stretched into thin sheets reflects heat. The rocket engines of the first American space shuttle were lined with gold to reflect heat.

The chief gold-producing country today is South Africa. This country has the richest gold field in the world.

More recently, prospectors found gold along the banks of the Amazon River in Brazil.

The element gold is more important in our lives than most people realize. Its physical properties make it useful and valued the world over. Maybe you will find a chunk of gold someday—or maybe you will invent a new use for this element!

Naming Elements

Elements have been given symbols so that people have a short, easy way to write their names. The symbol for an element may be one or two letters. Look at the element names on this page. Match them with the symbols on page 45. It is easy for us to see the relationship between these symbols and the elements' names. However, some symbols are taken from the Latin words for the elements. For example, the Latin word for iron is *ferrum,* so the symbol for iron is Fe. The Latin word for gold is *aurum.* Can you guess what the symbol for gold is? We have seen that a standardized system of measure makes it easy for people all over the world to share information by using common units. The same is true for elements. People all over the world use the same symbols for elements to be able to quickly and accurately pass along information.

Oxygen

Hydrogen

Carbon

Nitrogen

Helium

Neon

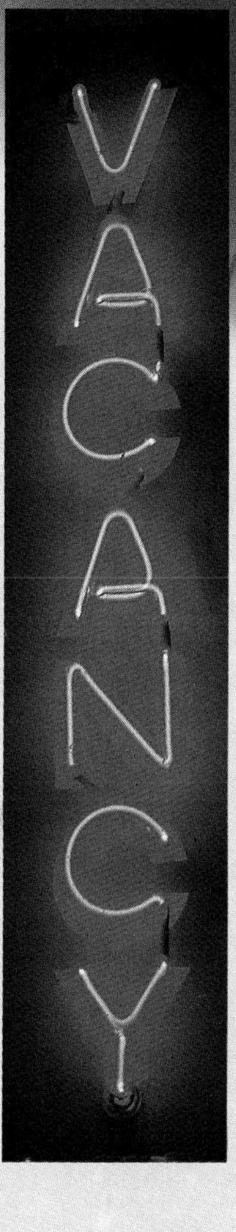

Ne He N H O C

TRY THIS **Activity!**

Where Do Element Names Come From?

What You Need

reference books, *Activity Log* page 19

You know that Latin words are sometimes used when naming elements. Other methods have been used to name elements too. Hypothesize where you think the names of the elements came from on Activity Log page 19. Write your hypothesis in the chart. Use reference books to check your answer and write any corrections in the "Research Result" column. If you discovered a new element, what would you name it? What would its symbol be?

Sometimes the element symbol is made of letters that are not at the beginning of the element name. Do the Try This Activity to learn the origin of element names and see examples of symbols that do not use the first one or two letters of the element name.

The Elements in Our Bodies

Health Link

Our bodies are also made of elements; after all, we are matter too. Can you think of any elements our bodies contain? Record your answers on page 20 of your ***Activity Log.*** The most common element in our bodies is oxygen, which is a part of water.

How do you think our bodies replace and add to these elements as we grow? Why do cereal boxes and other food packages list some of these elements as ingredients? What elements can you find listed on food package labels? Bring in two or three ingredient lists from food packages. Compare lists with your classmates. Which elements are not mentioned?

Minds On! Some elements, such as gold and uranium, are considered more valuable than other elements, such as iron, tin, or oxygen, because they are so rare. In your ***Activity Log*** on page 21, list three elements you think are valuable. Compare your list with others. Discuss what makes something valuable.

65% oxygen

20% carbon

15% other elements—hydrogen, nitrogen, calcium, phosphorus, potassium, sulfur, sodium, chlorine, magnesium, iodine, iron, etc.

Even gold, which is considered one of the most valuable elements on Earth, isn't good for everything. Have you ever read the story of King Midas? He was given the power to turn objects to gold by just touching them. But then everything he touched turned to gold. King Midas turned his eyeglasses to gold. How did the physical properties of his eyeglasses change when they turned to gold? What properties made them less useful than before? What are some other things that would be much less useful if they were made of gold?

Sum It Up

An atom is the smallest piece of an element that still has the properties of that element. You know this now because you've learned that everything in the world is made of tiny building blocks called atoms. Some things, like you, are made of different kinds of atoms. In an element like gold, though, all the atoms are the same. Would you be able to cut a piece of gold wire into atoms? Not in the classroom, that's for sure!

Critical Thinking

1. How are bricks and bananas alike? What makes them different?

2. What do you think would make one element more valuable than another?

3. In what ways would using symbols for elements help you communicate detailed research information to a person who does not speak or read English?

Atoms on the Move!

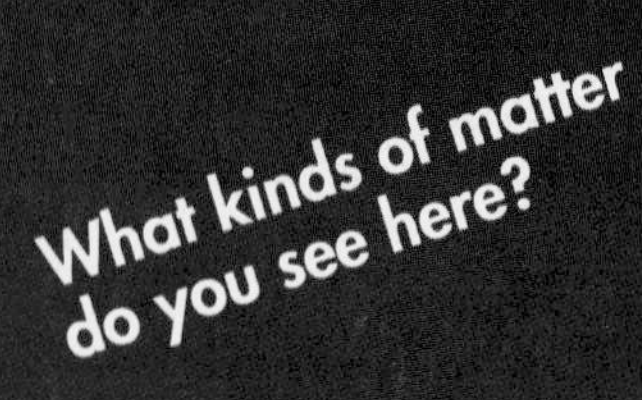

What kinds of matter do you see here?

What does ice become when it melts?

Where does water go when it evaporates?

You know that matter is made of atoms. Every day you see matter in three physical states—solid, liquid, and gas. There are three physical states of matter in your body as well. Your bones and muscles are solids, your blood and digestive juices are liquids, and the oxygen and carbon dioxide in your lungs and blood are gases.

You may have watched matter change from one state to another. This is called a physical change. During a physical change, the matter itself doesn't change.

A puddle of water drying up after a rainstorm and icicles forming in winter are examples of changes of state. What other examples can you think of?

How much space do solids, liquids, and gases take up? Do solids, liquids, and gases have a shape of their own? Do the Explore Activity on pages 50–51 to find out.

TRY THIS **Activity!**

A Change of State

What You Need

cup of very warm water, ice cubes, *Activity Log* page 22

Let's do an activity to observe changes in matter. Put three ice cubes in a cup of very warm water. Predict what you think will happen. Write your predictions in your ***Activity Log.*** Observe the ice and water. Did your observations support your prediction?

EXPLORE

Activity!

Matter Takes Shape!

In this activity you are going to investigate some properties of a solid, a liquid, and a gas.

What You Need

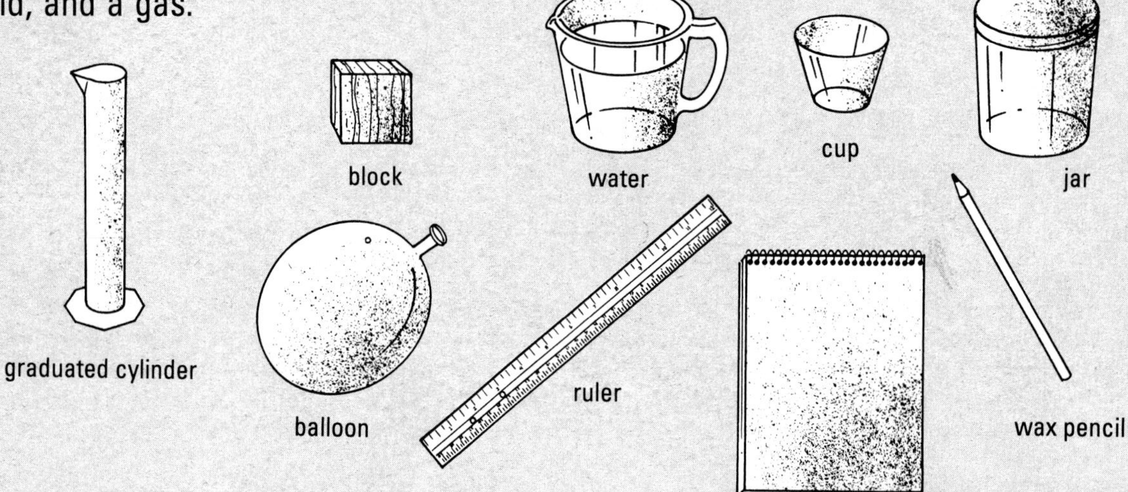

Activity Log page 23-24

What To Do

1 Using the graduated cylinder to measure, pour about 50 mL of water into the taller container, as shown in the picture. What shape does the water take? With the wax pencil, make a mark on the container at the level of the top of the water.

2 Now, pour the water into the shorter container. What happened to the shape of the water? With the wax pencil, make a mark at the top of the water again. Has the volume of the water changed? How could you find out? Measure the volume of the water.

3 Look at the solid block. What shape is it? Measure its length and width. Put the block into an empty container. Does the shape of the block change? Explain.

4 Blow up the balloon halfway, and tie the end or have a group member hold it so that the air does not escape. What is the shape of the air? Gently squeeze the balloon. Are you able to change the shape of the air?

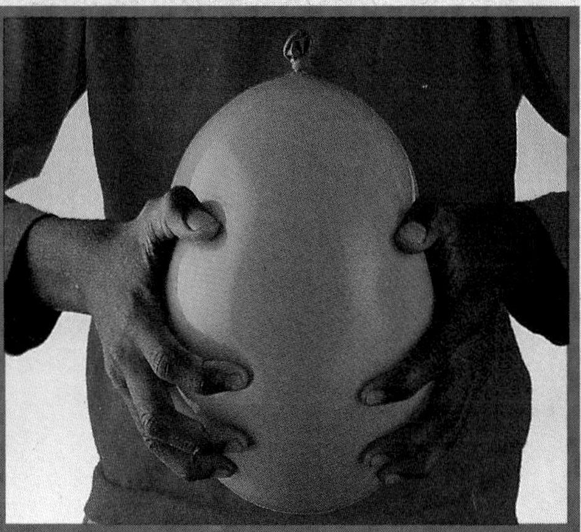

What Happened?

1. What is the physical state of each of these?
   a. water **b.** block **c.** air
2. Think about the block. What is the shape of the block? How did the size and shape of the block change?
3. Does every solid have its own definite size and shape?
4. Think about the activity you just did and complete the chart in your ***Activity Log.***

What Now?

1. What are some properties of solids, liquids, and gases that you found out about in this activity?
2. What shape does air take when you fill a basketball? A football?
3. Would you be able to fill a basketball and a football with the same solid? Why? What about with water?
4. List five types of matter in the chart in your ***Activity Log.*** Does each type of matter you listed exist as a solid, a liquid, or a gas? Make a check mark in the columns that apply. Compare your chart with a classmate's chart.

Solid, Liquid, Gas

As you learned in the Explore Activity, a **solid** has a fixed shape and a fixed volume, and a **liquid** has a fixed volume but not a fixed shape. A **gas** doesn't have a fixed volume or shape; it expands to fit whatever space is available.

But what keeps your pencil solid? Why does milk flow all over the table when it spills rather than stay in the shape of the glass? How can the same particles of matter change from a solid to a liquid to a gas and back to a liquid again?

Here's the answer: all particles of matter are constantly moving. The amount of heat energy they have determines whether the matter is a solid, a liquid, or a gas. If the particles of matter start moving faster or slower, the physical state of the matter may change. Though matter may change, it is never created or destroyed.

The particles in gases are far apart. They have the most heat energy, which is why they move so fast. Most gases must be kept in containers or their quickly moving particles will keep spreading and spreading.

The particles in solids are packed tightly together in a definite pattern, which gives the matter a definite size and shape. The particles don't have much room to move, but they still vibrate.

The particles in liquids slip and slide into any open areas they find. Not only do the particles move more freely, they also have more energy than the particles in a solid and move more quickly.

Minds On! Imagine that you are with three classmates, standing as close together as possible, with two in front of the other two. Hold hands and march slowly in place. While you march, stay in the same positions. You represent particles in a solid. Particles in a solid remain in a definite pattern at all times. Next, imagine that the four of you move apart slightly and drop hands. March a little faster and move around, but don't move too far away from each other. You represent particles in a liquid. By dropping hands but not moving far apart, you show that particles in liquids are not in a definite pattern but still remain in a certain area. Now imagine that you can race around the classroom as fast as you can, crashing into each other and the walls. You now represent particles in a gas! As matter changes from solid to liquid to gas, each state has more and more energy.●

Do the Try This Activity on this page to demonstrate atoms moving in a liquid.

TRY THIS Activity!

A Moving Activity

What You Need

2 clear plastic cups, hot water, cold water, blue food coloring, dropper, *Activity Log* page 25

If you put food coloring into water and did not stir, would it sink, float, or begin to mix with the water? Try it! Fill a plastic cup with cold water. When the water is completely still, gently put in two drops of food coloring. Do not stir, and observe what happens. Now try the same thing using hot water. What differences in the movement of the food coloring did you observe? Using your observations and information from the text, infer what caused the differences. Record your results in your ***Activity Log.***

Evaporation

Have you ever spilled a drink on yourself and found it was dry by the time you got home? Where did the liquid go? The change of a liquid into a gas is called evaporation (i vap′ ə rā′ shən). But what caused the liquid to evaporate? Do all liquids evaporate at the same rate?

Although different kinds of matter are liquids, they do not evaporate at the same rate. The different properties of the matter will cause differences in the rate of evaporation. The rate of evaporation is also affected by the environment. Can you think of a way to speed up evaporation?

Minds On! Imagine that you left an ice cube on the kitchen counter for an hour. What would you expect to happen? Now, imagine that you had left the ice cube there for a whole week. What would you expect to find at the end of the week? On page 26 in your ***Activity Log,*** explain what would happen in both cases. ●

Why can these children smell the freshly baked bread?

Where does the water go that the window washer leaves behind?

When energy is added to matter, the matter's particles move faster. When the particles of matter move faster, they also move farther apart. This means that as matter gets more energy, it expands, or gets larger. As it cools and its particles move more slowly, it contracts, or gets smaller.

As the temperature goes up in a thermometer, the liquid in the tube expands and rises. When the temperature cools, the liquid contracts and the level falls.

Do the Try This Activity on this page to see how heat affects the evaporation of a drop of water. It takes energy for a liquid to change to a gas, or evaporate. Since your finger is hotter than your desk, the transfer of the heat energy was faster from your finger to the water than from your desk to the water. The water evaporated faster from your finger.

TRY THIS **Activity!**

The Furnace in Your Finger

What You Need

dropper, plastic cup with some water, *Activity Log* page 27

Which do you think would evaporate faster, a drop of water on your desk or a drop of water on your finger? Write your prediction in your ***Activity Log.*** Now use a dropper to put a drop of water on your desk and one on your finger. Spread them both out. Observe the drops. Which one evaporated faster? Was your prediction correct? Draw a cartoon or write a description of how the evaporation of water made your finger feel.

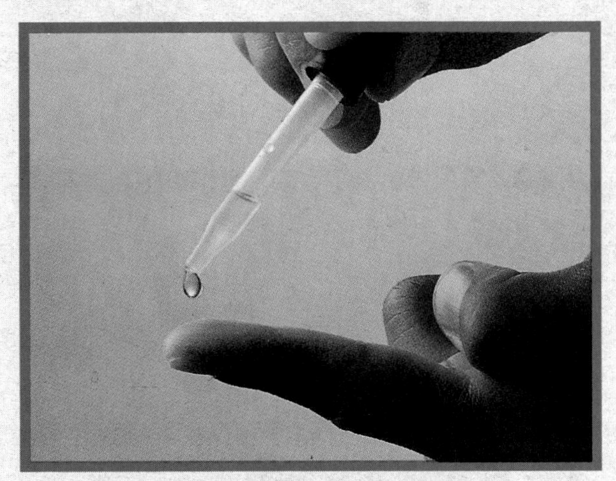

Remember the ice cubes we pretended to leave on the kitchen counter? What changes in physical state occurred? How did the changes happen? Check your answers on page 26 in your ***Activity Log*** and make changes or additions, if needed.

What about the ice cube that sat on the counter for a week? Did the ice cube disappear completely? Explain what happened to cause these physical changes.

Ice changing physical state

Sweating to keep cool

Cool It! Health Link

When our body temperature rises and our bodies send water to the surface of the skin, we sweat! The heat from our bodies causes the water to evaporate. The water takes heat energy from our bodies as it changes into water vapor. This helps us feel cooler. In the same way, if we get wet on a winter day, the evaporating water takes heat from our bodies, making us feel uncomfortably cold.

Most matter will **expand,** or have a larger volume, when heated and **contract,** or have a smaller volume, when cooled. Have you ever noticed that sidewalks are made of rectangles with spaces between them? Why aren't sidewalks made in one long strip all the way down the street? What might happen to a long strip of concrete on a very hot summer day? Would a similar thing happen to concrete in rectangles with space between them?

Buildings and bridges expand with heat and contract with cold. Strips are put in buildings and bridges to prevent this movement from causing damage to the structure.

The Exception to the Rule

Have you ever put water in an ice cube tray and discovered the level was higher when the water froze? The ice was taking up more room than the water. But didn't we just say that most matter expands when heated and contracts when cooled? The exception to this "rule" is water. Water expands when heated and contracts when cooled until it gets near its freezing point (4°C). Then it expands! Potholes in streets are examples of damage done by water expanding into ice.

You can see evidence of frozen water expanding by doing the Try This Activity on this page.

Sugaring Time Literature Link

Have you ever walked through deep, crusty snow or tried to get across an icy patch? Jumping Jack and Tommy aren't too excited about trudging through the ice and snow to get to the sugarbush. As you read *Sugaring Time,* you will see that ice is a factor to be considered when the sap's rising.

TRY THIS Activity!

Pop Its Top!

What You Need

35-mm film canister, water, *Activity Log* page 28

Fill the canister with water and carefully place the lid on it. Put it in the freezer. Check it in 5 minutes. Did you observe any difference in the canister? Check it again in 2 hours. What was the first thing you observed when you saw the canister with the water frozen inside?

Potholes start in the winter when rain gets into a crack in the road. As the water freezes, it expands, making the crack bigger. More water flows into the bigger crack. Soon the water freezes and expands, forcing the crack to open wider. In this way a little crack can become a large, dangerous hole.

Sum It Up

The properties of matter are not determined just by the kinds of atoms it's made of. What matter feels like and how it looks are also determined by its physical state. Matter commonly exists in three physical states: solids like your bed or your socks, liquids like milk or water, and gases like the air you breathe.

Properties of matter can change if heat is added or removed. Adding heat speeds up the movement of atoms. As the atoms move faster, the matter may change from a solid to a liquid to a gas. Taking heat away slows the movement of atoms and makes them move close together. For example, as the temperature falls, water vapor in the air can change to beads of water on the roof and then to solid icicles hanging from the edge.

Critical Thinking

1. Blow up a balloon. What would happen if you put the balloon in the sun? What would happen if you put the balloon in a refrigerator? Try it to see if your predictions are correct.

2. Would a metal bridge be longer in the summer or in the winter? Why?

3. If you live in an area with very cold winters, should water be kept in your outdoor community swimming pools all winter? Why or why not?

Theme T SYSTEMS and INTERACTIONS

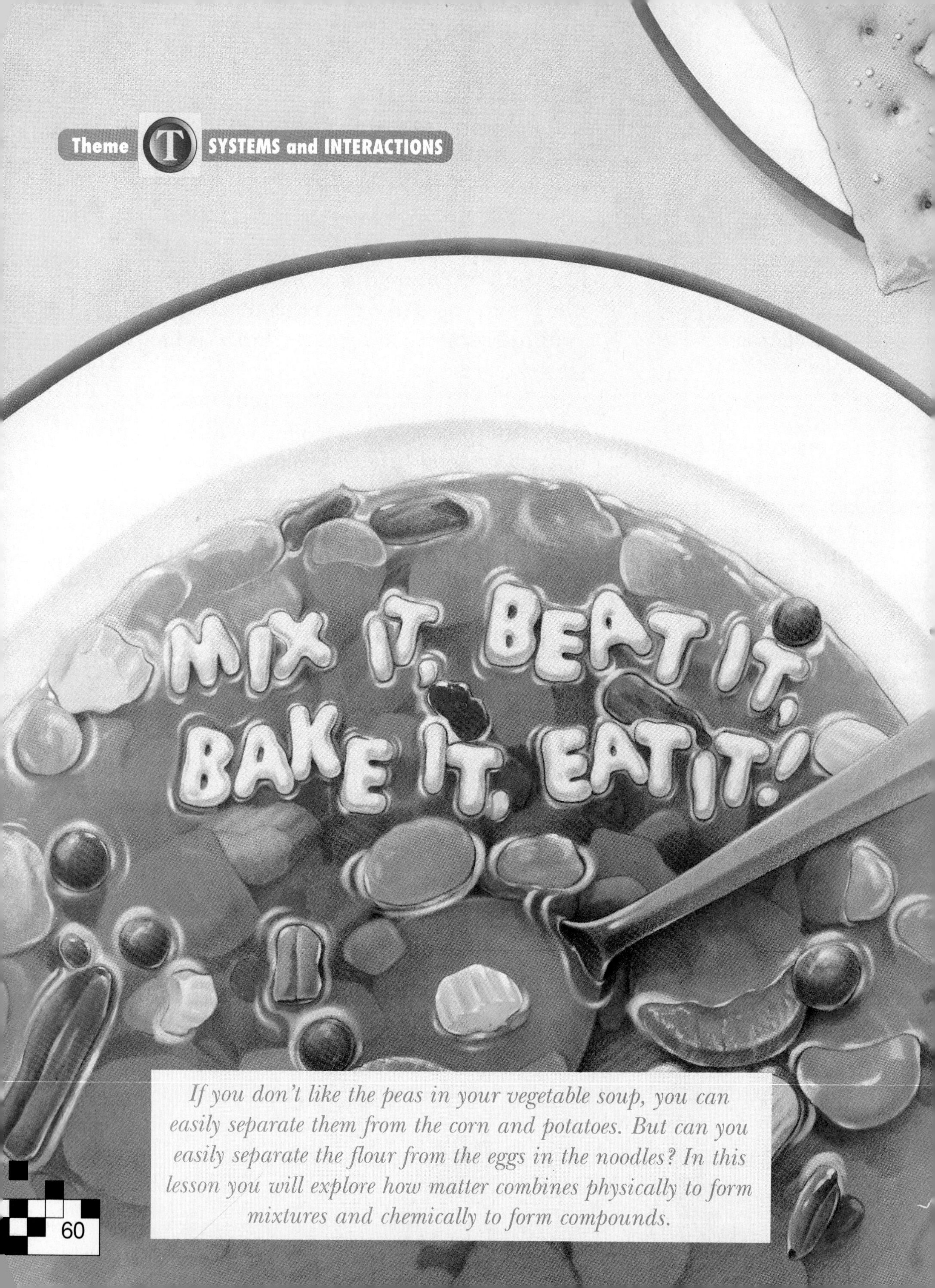
MIX IT, BEAT IT,
BAKE IT, EAT IT!

If you don't like the peas in your vegetable soup, you can easily separate them from the corn and potatoes. But can you easily separate the flour from the eggs in the noodles? In this lesson you will explore how matter combines physically to form mixtures and chemically to form compounds.

What's your favorite mixture? Spaghetti and meatballs, root beer and ice cream, or dirt and water in the backyard? Matter can be mixed together in an unlimited number of ways. Some people think mixtures are just solids in liquid, like drink crystals in water. But here are some other examples of mixtures:

- gases (the oxygen and carbon dioxide you breathe)
- liquids (lemon juice and water)
- solids (fruit cocktail)
- a gas and a liquid (soft drinks)
- a gas and a solid (air with dust in it)

Minds On! The toppings on a pizza form a mixture. If you had a pizza with everything, what would you separate? The olives? The mushrooms? How would you identify the kinds of matter you wanted to separate? Do you know any other ways to separate mixtures? ●

EXPLORE

Activity!

If You Put It Together, Can You Take It Apart?

Matter can form mixtures. Mixtures can be separated by using their physical properties. You already thought about one way to separate mixtures with the pizza. In this activity you are going to use two other ways to separate some mixtures.

What You Need

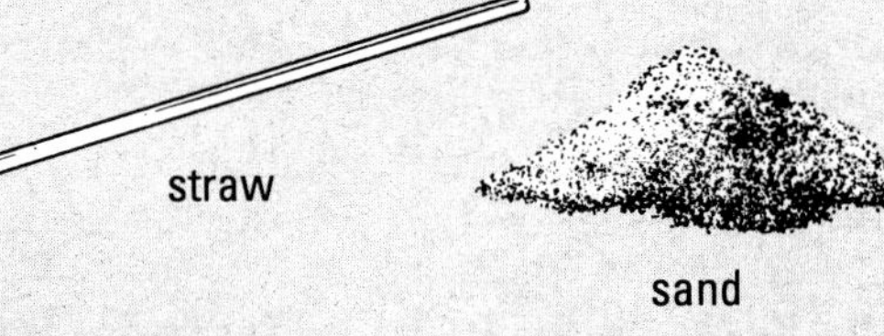

cup

filter

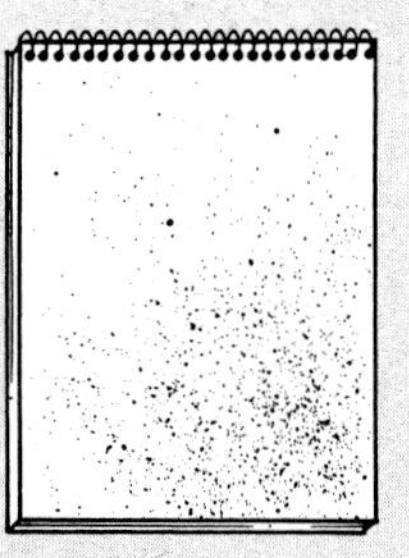

Activity Log page 29-30

What To Do

1 Add about one spoonful of salt to about one half cup of warm water and stir. What happens to the salt in the water? Does it disappear? Put the straw straight down about 2 cm into the water and put your finger over the top to hold the liquid in. Take the straw out and release a few drops of the liquid onto your tongue. How does it taste? How does this prove that the salt did not disappear?

2 In a second cup, add a spoonful of sand to a half cup of warm water and stir. What happens to the sand in the water? Does it disappear?

3 How can you separate the two combinations of matter you have made? Talk to the members of your group about your ideas before you continue.

4 Put a filter over one of the empty cups as shown in the picture. Slowly pour the sand and water through the filter and allow it to drip into the bottom cup. What happens to the sand in the water?

5 Try to filter the salt water in the same way. What happens? Taste the water using the straw in the same way as in step 1.

6 Pour out most of the salt water. Leave a small amount in the cup.

7 Look at your salt water every day for a few days. One member of your group should record the group's observations. What is left in the cup after the water is gone?

What Happened?

1. How were you able to separate the water and the sand?
2. How were you able to separate the water and the salt?
3. Could you use evaporation to separate the sand and the water? Plan a way and try it.

What Now?

1. Filtering is used to separate many mixtures. What physical property is used to separate some mixtures with a filter?
2. Was the salt water a mixture? How do you know? HINT: How were you able to separate it?
3. Hypothesize as to how you could separate a mixture of white sand and salt. Test your hypothesis.

Mixtures

In the Explore Activity, you made two different mixtures—sand plus water and salt plus water. In a **mixture** (miks′ chər), two or more kinds of matter are combined, but they keep their original properties. The different kinds of matter can be separated again using the physical properties of the matter. Physical properties include size, shape, volume, density, and physical state. Matter can have changes in its physical state, but it is still the same kind of matter. For example, ice is the same kind of matter as water, but it has different physical properties because it is in a different physical state. Size is another physical property of matter that can be changed without changing the kind of matter.

You could use the physical properties of color and size to separate a mixture of marbles. (You could pick out the large cat eyes!)

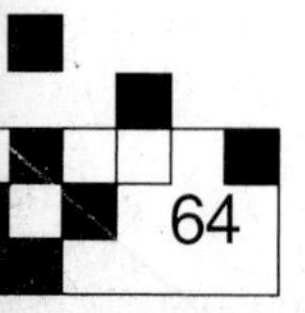

In the Explore Activity you used two ways to separate matter in a mixture—filtration and evaporation.

Filtration (fil trā′ shən) separated the sand from the water. One of the sand's physical properties (size) kept it from going through the filter. Coffee filters work the same way, keeping coffee grounds out of the coffee. You may have other kinds of filters at home, such as sieves and colanders. They have bigger holes, but they function the same way.

Evaporation separated the salt from the water. The physical properties of water made this possible. When the water changed from a liquid to a gas, the salt was left behind. In the same way, if you spill a sugary drink on the kitchen counter at home, you may notice a sticky area after the liquid evaporates. When the liquid changes to a gas, it leaves a sticky spot of sugar behind.

Sugaring Time Literature Link

As you read this book, look for the ways evaporation and filtration are used in the making of maple syrup.

Compounds

Nearly all atoms combine with other atoms. Two or more atoms joined together form a **compound** (kom′ pound). Usually, a chemical change takes place in the elements when a compound is formed. The properties of a compound are different from the properties of the elements that make it up. For example, at room temperature hydrogen and oxygen are gases, but when they combine, they form liquid water. The elements that make up a compound can't be separated by using their physical properties. Compounds can only be broken down by using chemical properties.

Electrical charges hold the atoms in a compound together. Compounds in which electrical charges are shared between atoms are called **molecules** (mol′ ə kūls′). Some molecules are made of two or more of the same kind of atom. Other molecules are made of different kinds of atoms. A molecule of water is made up of two atoms of hydrogen and one atom of oxygen.

Water, which can be used to put out fires, is a compound of two flammable gases.

TRY THIS Activity!

A Different Use for Lemon Juice

What You Need

lemon juice (fresh or concentrated), dropper, tarnished penny, tissue, *Activity Log* page 31

You can see tarnish on a penny. A chemical change made the tarnish. Another chemical change can take the tarnish away. Put a drop of lemon juice on the penny. Let it sit, without moving it, for 5 to 10 minutes and then wipe off the juice. The citric acid in the lemon juice has chemically removed the tarnish!

Tarnish is a compound you might have seen—you might even have some in your pocket! Tarnish forms when air reacts with certain metals, making them look dull rather than shiny. Do the activity on this page to remove tarnish an easy way.

TRY THIS Activity!

Can Wool Rust?

What You Need

3 cups, 3 pieces of steel wool, water, cooking oil, *Activity Log* page 32

What does it take to make rust? Try this activity to find out. Dip one piece of steel wool in water. Dip the second one in cooking oil and then in water. Do nothing with the third piece. Put all three someplace where they won't be disturbed. Predict which pieces of steel wool will have rust on them after two days. Observe any changes. What does this activity tell you about the chemical change that causes the compound rust? What do your findings tell you about possible ways to prevent or slow down the formation of rust? Think of other ways to prevent the steel wool from rusting. Test your predictions.

Another chemical change, or compound, you might be familiar with is rust. When oxygen and moisture interact with iron or steel, rust forms. Rust eats away, or corrodes, the iron or steel on which it forms, weakening it. The physical properties of rust are not the same as the properties of the oxygen, moisture, or metal. Can you think of a way rust might be prevented? Do the Try This Activity on this page to prevent rust.

Focus on Technology

Getting Rid of Rust

Bridges throughout the country are being worn away by the salt spread on them in winter to melt snow and ice. Bridges over salt water are also splashed with more salt from below. About half the bridges in the United States—250,000 of them—have been weakened in this way and may not be safe for heavy loads.

Here's what happens. The bridges are supported by concrete with steel bars embedded in it. Salty water seeps into the concrete and combines with the steel to form the compound rust. Rust expands the bars, putting pressure on the surrounding concrete. Cracks form, allowing more salty water to reach the bars and causing more rust.

Stopping rust with electricity

Scientists have learned that a small amount of electricity stops—or at least slows—the chemical change that causes rust. To stop a bridge from corroding, workers cover the bridge with wire mesh and run electricity through the mesh from nearby light poles. Then they cover the mesh with more concrete.

Does this waste a lot of electricity? Not at all! Protecting a medium-sized bridge from rust requires about the same amount of electricity as running a small space heater.

Kitchen Mixtures

We use—and form—mixtures and compounds every day. One evening we might eat a mixture of spaghetti and meatballs for dinner. You know this is a mixture because the two kinds of matter, the spaghetti and the meatballs, can be separated using their physical properties.

What physical properties would you use to separate this mixture?

Combine some beef, onion, garlic, tomato sauce, salt, and pepper, put it into a taco shell, top it with cheese and lettuce, and you have a mixture that you'd probably rather eat than take apart!

Some kinds of matter cannot be mixed easily. Mixing vinegar and oil for salad dressing isn't easy. They separate very quickly

Sum It Up

We have access to elements that exist on Earth. Their combinations in mixtures and compounds have led to amazing products and conveniences. We use, and form, mixtures and compounds every day. When you form a mixture, you can still separate the matter by physical properties. You discovered that filtration and evaporation are two ways to separate a mixture.

Critical Thinking

1. Is salad a compound or a mixture? How do you know?

2. Is ice a compound or a mixture? How do you know?

3. Suppose you have a mixture of iron filings and salt. How could you separate them?

The fact that water and oil don't mix easily is good news when it comes to oil spills in the ocean. The oil stays on the surface of the water, where workers have a better chance of skimming it off.

What's the Matter?

Now you know what's the matter—every living and nonliving thing in the world is matter! The world seems enormous, but all of it—even mountains, giant redwoods, the oceans, and you—can be broken down into smaller and smaller bits of matter. We can't even see the smallest bits, called atoms, without a special microscope.

Of course, mountains, redwoods, the oceans, and you are each made of different kinds of matter. We can tell mountains from redwoods by observing properties such as size, shape, color, and roughness. We can even tell taller mountains from shorter mountains because we know how to measure properties of matter.

Ocean

Redwood

Gold

Only 109 different kinds of matter, called elements, make up everything in the world. We've learned to use the unique physical properties of each element in different ways. The warm glow of gold makes beautiful jewelry. Radioactive uranium is vital in producing atomic energy.

Mercury is often used in thermometers because it's sensitive to temperature. The physical and chemical properties of copper make it ideal for electrical wires.

You know, though, that some properties of matter change as its physical state changes. Adding heat causes the atoms in matter to move more quickly and therefore farther apart. In this way, solids can change to liquids, and liquids to gases.

You and every other living thing on Earth depend on this change in physical state.

The strength of iron supports buildings.

The lighter weight of aluminum is better for airplanes.

During evaporation, the liquid changes to a gas—water vapor. The water vapor cools in the air high above Earth, returning to its liquid state.

The sun speeds up the evaporation of water from oceans, lakes, and rivers.

Rain once again fills the lakes and rivers—and makes inviting puddles on the sidewalks!

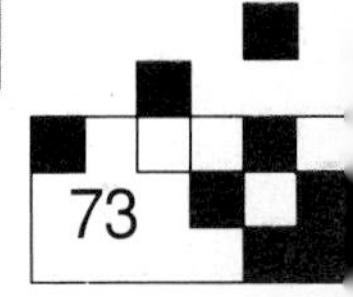

Removing heat from matter has the opposite effect on the matter's physical state. The atoms move more slowly and cluster closer together. A gas turns to a liquid, and a liquid to a solid.

Industries use this change in physical state when they inject melted plastic into molds and let it cool. As the heat moves out of the hot plastic into the cooler air, the atoms in the plastic move closer together. They take the shape of the mold. Unless the plastic is heated again, it will stay permanently in the shape of the mold.

Besides changing physical states, elements can combine in millions of ways. Some combine physically into mixtures and keep their own physical properties (as with the pizza). Others combine chemically into compounds. Then they often develop entirely different properties, as when the gases oxygen and hydrogen combine to form water.

Sodium, an explosive in water, and chlorine, a poisonous gas, combine to form common table salt.

Remember Roy Plunkett, the scientist whose experiment did not produce the results he expected? He had combined different kinds of matter to create a new compound. But the compound didn't have the physical properties he had planned on. He did notice, though, that this new compound was not affected by heat, cold, water, or other chemicals. What's more, its surface was smooth and slippery. He had invented Teflon™!

You may have pans at home that are coated with Teflon™ to keep food from sticking to them. Only 5 percent of Teflon™ is used on cookware. The rest helps us in a wide variety of ways.

Teflon™ frying pan

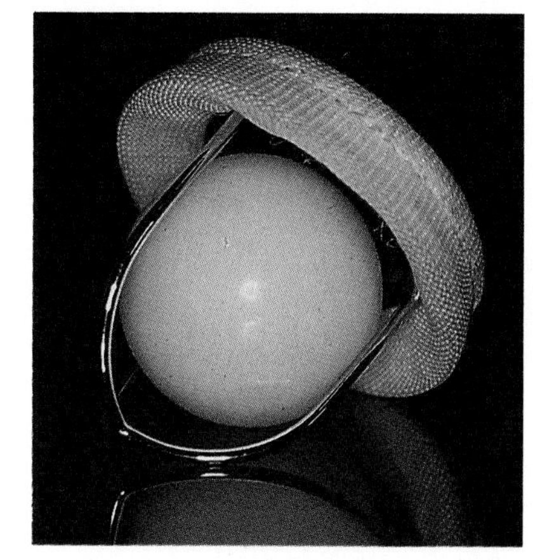

Teflon™ coats artificial heart valves so they slide smoothly.

It's amazing to realize that so few elements combine to produce everything we find in nature and everything scientists have ever invented. However, we must find better ways to take care of the tiny bits of matter that form the oceans, cities, and people of our world. From recycling plastics and other trash to preserving the rain forests, we have to use our knowledge of matter and how it interacts to help make Earth a clean and healthy place for all living things. We need to protect and treasure a world of so much wonder made from so few elements.

GLOSSARY

Use the pronunciation key below to help you decode, or read, the pronunciations.

Pronunciation Key

a	**a**t, b**a**d	d	**d**ear, so**d**a, ba**d**
ā	**a**pe, p**ai**n, d**ay**, br**ea**k	f	**f**ive, de**f**end, lea**f**, o**ff**, cou**gh**, ele**ph**ant
ä	f**a**ther, c**a**r, h**ea**rt	g	**g**ame, a**g**o, fo**g**, e**gg**
âr	c**are**, p**air**, b**ear**, th**eir**, wh**ere**	h	**h**at, a**h**ead
e	**e**nd, p**e**t, s**ai**d, h**ea**ven, fr**ie**nd	hw	**wh**ite, **wh**ether, **wh**ich
ē	**e**qual, m**e**, f**ee**t, t**ea**m, p**ie**ce, k**ey**	j	**j**oke, en**j**oy, **g**em, pa**g**e, e**dg**e
i	**i**t, b**i**g, **E**nglish, h**y**mn	k	**k**ite, ba**k**ery, see**k**, ta**ck**, **c**at
ī	**i**ce, f**i**ne, l**ie**, m**y**	l	**l**id, sai**l**or, fee**l**, ba**ll**, a**ll**ow
îr	**ear**, d**eer**, h**ere**, p**ier**ce	m	**m**an, fa**m**ily, drea**m**
o	**o**dd, h**o**t, w**a**tch	n	**n**ot, fi**n**al, pa**n**, k**n**ife
ō	**o**ld, **oa**t, t**oe**, l**ow**	ng	lo**ng**, si**ng**er, pi**n**k
ô	c**o**ffee, **a**ll, t**au**ght, l**aw**, f**ou**ght	p	**p**ail, re**p**air, soa**p**, ha**pp**y
ôr	**or**der, f**or**k, h**or**se, st**or**y, p**our**	r	**r**ide, pa**r**ent, wea**r**, mo**r**e, ma**rr**y
oi	**oi**l, t**oy**	s	**s**it, a**s**ide, pet**s**, **c**ent, pa**ss**
ou	**ou**t, n**ow**	sh	**sh**oe, wa**sh**er, fi**sh** mi**ssi**on, na**ti**on
u	**u**p, m**u**d, l**o**ve, d**ou**ble	t	**t**ag, pre**t**end, fa**t**, bu**tt**on, dress**ed**
ū	**u**se, m**u**le, c**ue**, f**eu**d, f**ew**	th	**th**in, pan**th**er, bo**th**
ü	r**u**le, tr**ue**, f**oo**d	t͟h	**th**is, mo**th**er, smoo**th**
u̇	p**u**t, w**oo**d, sh**ou**ld	v	**v**ery, fa**v**or, wa**v**e
ûr	b**ur**n, h**urr**y, t**er**m, b**ir**d, w**or**d,c**our**age	w	**w**et, **w**eather, re**w**ard
ə	**a**bout, tak**e**n, penc**i**l, lem**o**n, circ**u**s	y	**y**es, on**i**on
b	**b**at, a**b**ove, jo**b**	z	**z**oo, la**z**y, ja**zz**, ro**s**e, dog**s**, hous**es**
ch	**ch**in, su**ch**, ma**tch**	zh	vi**si**on, trea**s**ure, sei**z**ure

atom (at´əm) the smallest piece of an element that still has the properties of that element

compound (kom´pound) matter formed from two or more elements joined together

contract (kən trakt´) to draw parts closer together or make smaller

cubit (kū´bit) the distance from the elbow to the tip of the middle finger (approximately 18 inches)

density (den´si tē) the amount of mass a substance has for a certain volume

element (el´ə mənt) matter made of one kind of atom

evaporation (i vap´ə rā´shən) the change from a liquid to a gas

expand (ek spand´) to push parts farther apart or make larger

eyeprint (ī´print) the pattern of the blood vessels located on the retina

fathom (fath´əm) the distance between fingertips with the arms stretched out to the sides (6 feet)

filtration (fil trā´shən) the passing of matter through a filter to separate larger particles

fingerprint (fing´gər print´) an impression of the markings on the inner surface of the tip of a finger

foot (fu̇t) originally the length of the king's foot (12 inches)

gas (gas) matter that does not have a definite volume or a definite shape

hand (hand) the width of the palm (4 inches)

landfill (land´fil´) an area of land that has been filled in, usually by dumping refuse and mixing or covering it with soil

liquid (lik´wid) matter that has a definite volume but not a definite shape

mass (mas) the amount of matter in an object

matter (mat´ər) anything that takes up space and has mass

mixture (miks´chər) a combination of two or more different types of matter in which each type of matter keeps its own properties

molecule (mol´ə kūl´) a compound in which electrical charges are shared between atoms

newton (nü´tən) the unit for measuring weight

pace (pās) the distance of a single step

property (prop´ər tē) a characteristic of an object or substance that can be sensed or measured

recycle (rē sī´kəl) to change into another form or shape for reuse

rust (rust) the reddish brown or orange coating that forms on the surface of iron or steel that has been exposed to moisture and oxygen

solid (sol´id) matter that has a definite shape and a definite volume

span (span) the width of the outstretched hand from the tip of the thumb to the tip of the little finger (9 inches)

tarnish (tär´nish) a chemical change in metals exposed to oxygen

voiceprint (vois´print) a computer printout of the characteristics of the voice

volume (vol´ūm) the amount of space matter takes up

weight (wāt) a measure of the pull of gravity on an object

yard (yärd) the distance from the tip of the nose to the tip of the thumb of the outstretched hand (36 inches)

INDEX

CREDITS

Photo Credits:

Cover, The Image Bank/Dominique Sarraute; **1,** ©KS Studios; **2, 3,** Joe Sohm/The Stock Market, **3,** ©George Anderson; **6,** ©KS Studios/1991; **7,** (ti) ©Jonathan Eisenback/Phototake, (mi) ©Brian Parker/Tom Stack & Associates, (bi) ©Doug Martin, (r) ©Eric Meola/The Image Bank; **8,** ©Photo courtesy of Dupont; **8, 9,** The Bettmann Archive; **10, 11,** ©Studiohio; **12, 13,** ©KS Studios/1991; **14, 15,** ©Studiohio; **16,** ©Morton & White; **17,** (l) Margaret Cubberly/Phototake, (r) Photo courtesy of Eye Dentify Inc. (b) courtesy of The Ohio State University Dept. of Speech & Hearing; **18,** ©KS Studios/1991; **19,** (tl) (bl) ©Studiohio; (ml) (r) ©Doug Martin; **20,** (itl) (ibl) ©Doug Martin, (ir) courtesy Plastic Lumber Co.; **20, 21,** ©Rafael Macia/Photo Researchers; **22, 23,** ©KS Studios/1991; **24, 25,** ©Studiohio; **26,** ©Doug Martin; **26, 27,** Joe Sohm/The Stock Market; **28,** ©Studiohio; **29,** ©Doug Martin; **31,** ©Studiohio; **32, 33,** ©George C. Anderson; **34,** The Image Bank/Alvis Upitis; **35,** (t) The Image Bank/Garry Gay, (b) ©Ron Chapple/FPG International; **36, 37,** ©George C. Anderson; **38, 39,** ©Studiohio; **40,** ©Doug Martin; **41,** (l) ©Bruce Bishop/The Photo File, (r) ©Brent Turner/BLT Productions/1991; **42,** (itr) John de Visser/Masterfile, (iml) The Image Bank/Murray Alcosser, (ibl) ©Rick Altman/The Stock Market, (ibr) NASA; **42, 43,** IMTEK Engineering/Masterfile; **43,** (il) The Image Bank/Peter Frey, (ir) FPG International; **44, 45,** The Image Bank/Stephen Lynch; **45,** ©Andy Caulfield/The Image Bank; **46,** ©KS Studios/1991; **47,** ©Chas Krider; **48, 49,** ©Studiohio; **49,** (inset) ©Brent Turner/BLT Productions/1991; **50, 51,** ©Studiohio; **52, 53,** ©Peter Scoones/Planet Earth Pictures; **53,** ©Studiohio; **54,** ©Brent Turner/BLT Productions/1991/location courtesy of Katzinger's Delicatessen; **55,** (tl) ©Brent Turner/BLT Productions/1991; (mr) ©Studiohio, (br) The Image Bank/Francisco Ontanon; **56,** ©Fredrik Marsh; **57,** ©Bob Daemmrich; **58, 59,** ©Stephen Frisch/Stock Boston, Inc.; **62, 63, 64,** ©Studiohio; **65,** ©Doug Martin, (i) ©Studiohio; **66,** ©Superstock; **67,** (t) ©KS Studios/1991, (b) ©Doug Martin; **68,** (l) ©Studiohio; **68, 69,** ©Joe Sohm/UNIPHOTO; **70,** ©Morton & White; **71,** ©Michelle Barnes/Gamma Liaison, Inc.; **72,** (tr) ©Robin Smith/Tony Stone Worldwide/Chicago Ltd., (bl) ©Carolina Biological Supply, (br) The Image Bank/Jean Pierre Pieuchot; **73,** (t) ©Jan Staller/The Stock Market, (m) ©Tony Stone Worldwide/Chicago Ltd., ©Carl Roessler/FPG International; **74,** (t) ©Gary Benson, (b) ©Fredrik Marsh; **75,** (l) ©Al Paglialunga/Phototake; (r) ©Fredrik Marsh

Illustration Credits:

14, 24, 38, 50, 62, Bob Giuliani; **27,** Susan Johnston Carlson; **30, 31,** Jim Theodore; **60, 61,** Jack Graber